Data and Measurement Bingo Book

COMPLETE BINGO GAME IN A BOOK

U.S. Customary Units to Metric Units

inches to millimeters	multiply by 25
feet to centimeters	multiply by 30
yards to meters	multiply by 0.9
miles to kilometers	multiply by 1.6
fluid ounces to milliliters	multiply by 30
pints to liters	multiply by 0.47
quarts to liters	multiply by 0.95
gallons to liters	multiply by 3.8
to g	multiply by 28
grams	multiply by 0.45

Written By Rebecca Stark

ISBN 978-0-87386-461-9

Educational Books 'n' Bingo

Printed in the U.S.A.

DATA & MEASUREMENT BINGO DIRECTIONS

INCLUDED:

List of Terms

Templates for Additional Terms and Clues

2 Clues per Term

30 Unique Bingo Cards

Markers

1. **Either cut apart the book or make copies of ALL the sheets. You might want to make an extra copy of the clue sheets to use for introduction and review. Keep the sheets in an envelope for easy reuse.**

2. Cut apart the call cards with terms and clues.

3. Pass out one bingo card per student. There are enough for a class of 30.

4. Pass out markers. You may cut apart the markers included in this book or use any other small items of your choice.

5. Decide whether or not you will require the entire card to be filled. Requiring the entire card to be filled provides a better review. However, if you have a short time to fill, you may prefer to have them do the just the border or some other format. Tell the class before you begin what is required.

6. There are 50 terms. Read the list before you begin. If there are any terms that have not been covered in class, you may want to read to the students the term and clues before you begin.

7. There is a blank space in the middle of each card. You can instruct the students to use it as a free space or you can write in answers to cover terms not included. Of course, in this case you would create your own clues. (Templates provided.)

8. Shuffle the cards and place them in a pile. Two or three clues are provided for each term. If you plan to play the game with the same group more than once, you might want to choose a different clue for each game. If not, you may choose to use more than one clue.

9. Be sure to keep the cards you have used for the present game in a separate pile. When a student calls, "Bingo," he or she will have to verify that the correct answers are on his or her card AND that the markers were placed in response to the proper questions. Pull out the cards that are on the student's card keeping them in the order they were used in the game. Read each clue as it was given and ask the student to identify the correct answer from his or her card.

10. If the student has the correct answers on the card AND has shown that they were marked in response to the *correct questions,* then that student is the winner and the game is over. If the student does not have the correct answers on the card OR he or she marked the answers in response to *the wrong questions,* then the game continues until there is a proper winner.

11. If you want to play again, reshuffle the cards and begin again.

Have fun!

TERMS INCLUDED

28

32 FLUID OUNCES

55

AREA

BAR GRAPH

CAPACITY

CELSIUS

CENTURY

CIRCLE GRAPH

COORDINATES

CUBIC FOOT (FEET)

CUSTOMARY SYSTEM

DATA

FAHRENHEIT

FREQUENCY

GALLON

GRAPH

HECTARE(S)

HISTOGRAM

HOUR

KILOGRAM

KILOMETER

LINE GRAPH

LINE PLOT

LINEAR MEASUREMENT

LITER

MEAN

MEDIAN

METER

METRIC SYSTEM

MILE

MODE

PARALLEL

PERPENDICULAR

PLANE

POUND(S)

RANGE

SCALE

SQUARE FOOT (FEET)

STEM-AND-LEAF PLOT

TABLESPOON

TABLE

TALLY

TEMPERATURE

TIME

VARIABLE(S)

VOLUME

WEIGHT

X-AXIS

Y-AXIS

Additional Terms

Choose as many additional terms as you would like and write them in the squares.
Repeat each as desired.
Cut out the squares and randomly distribute them to the class.
Instruct the students to place their square on the center space of their card.

Data and Measurement Bingo

Clues for Additional Terms

Write three clues for each of your terms.

_____	_____
1.	1.
2.	2.
3.	3.

_____	_____
1.	1.
2.	2.
3.	3.

_____	_____
1.	1.
2.	2.
3.	3.

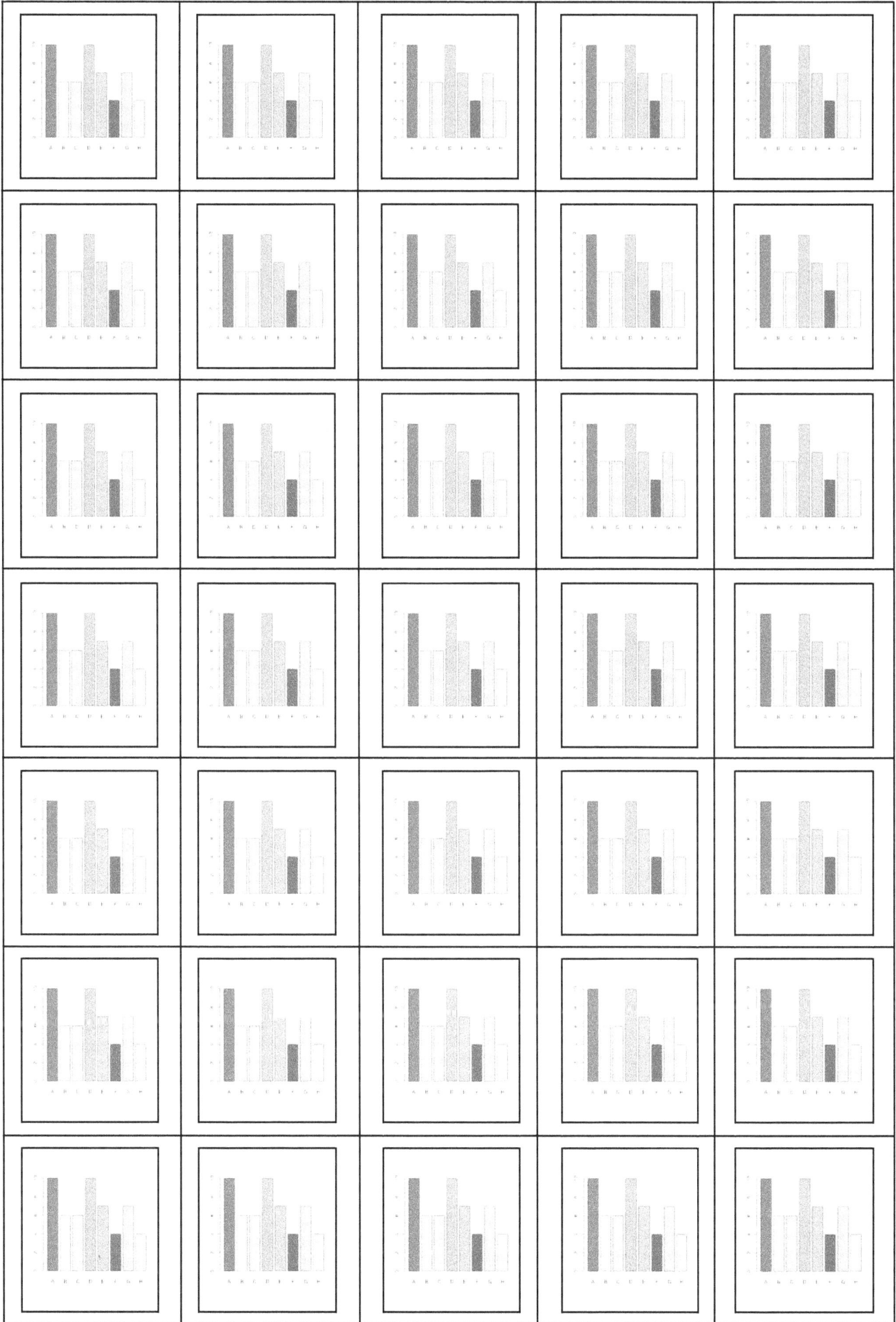

28 1. Number of pints in 14 quarts. 2. Number of inches in 2 1/3 feet. 3. Number of cups in 14 pints.	**32 Fluid Ounces** 1. Equivalent of 1 quart. 2. Equivalent of 8 pints. 3. Equivalent of 1/4 gallon.
55 1. The median in this series of scores: 46, 48, 54, 54, 55, 56, 56, 56, 60. 2. The mean of this series of scores: 40, 45, 55, 60, 60, 60, 65. 3. The range of these scores: 40, 60, 80, 80, 83, 91, 95.	**Area** 1. The measure of square units within a figure. 2. To find the surface ___ of a cube, use the formula $A = 6s^2$ with s being the length of each side of the cube. 3. To find the ___ of a triangle, use the formula $A = 1/2\ bh$, where b is the base and h is the height.
Bar Graph 1. A graph that uses the height or length of the rectangles to compare data. 2. A double one uses two rectangular shapes to show 2 sets of related data. 3. It displays data with vertical or horizontal bars.	**Capacity** 1. Amount of liquid a container can hold. 2. Pints, quarts and gallons are U.S. customary units of ___. 3. Milliliters, and liters are metric units of ___.
Celsius 1. On this scale under normal conditions water freezes at 0°. 2. On this scale under normal conditions water boils at 100°. 3. To convert from Fahrenheit to this scale, subtract 32 and multiply by 5/9.	**Century** 1. A unit of 100 years. 2. Equivalent of 10 decades. 3. There are ten in a millennium.
Circle Graph 1. Also called a pie graph or pie chart. 2. A circular chart that is cut into sections by lines going through its center. 3. A graph used to show how parts make up a whole.	**Coordinates(s)** 1. In a 2-dimensional plane, they are the pairs of numbers that specify the location of a point. The x-___ is given first and is followed by the y-___. 2. The x-___ of a point is its distance from the vertical or y-axis. 3. The y-___ of a point is its distance from the vertical or x-axis.

Data and Measurement Bingo

Cubic Foot (Feet) 1. There are 1,728 cubic inches in one. 2. There are 27 of them in a cubic yard. 3. There are 3 cubic yards in 81 of them.	**Customary System** 1. It is also called the English System. 2. This system is based on the foot as a unit of length, the pound as a unit mass, and the second as a unit of time, and the pint as a measure of volume or capacity. 3. This system uses non-metric units of measurement.
Data 1. Factual information such as measurements or statistics. 2. Sometimes these pieces of information are arranged in a table with rows and columns. 3. Sometimes these pieces of information are arranged by taking a tally.	**Fahrenheit** 1. On this scale under normal conditions water freezes at 32°. 2. On this scale under normal conditions water boils at 212°. 3. To convert from Fahrenheit to this scale, multiply by 9/5 and add 32.
Frequency 1. The number of times a value occurs in a set of data. 2. If you want to show the number of times a particular data value occurs in a given period of time, record the data in a ___ table. 3. Histograms are sometimes used to show this.	**Gallon** 1. 16 cups. 2. 8 pints. 3. 4 quarts.
Graph 1. A diagram that shows a relationship between two or more variables. 2. A diagram that compares the variations of one variable with one or more other variables. 3. Some different forms are bar, circle, and line.	**Hectare(s)** 1. Equivalent of 10,000 square meters. 2. There are 100 in a square kilometer. 3. It is longer than a square meter but smaller than a square kilometer.
Histogram 1. This type of graph uses vertical columns which to show frequency 2. A graphical display of tabulated frequencies. 2. This type of graph uses rectangles without gaps between them to show frequency.	**Hour** 1. 60 minutes. 2. 3,600 seconds. 3. 1/24 of a day.

Data and Measurement Bingo

Kilogram 1. 1,000 grams. 2. Its abbreviation is kg. 3. There are 1,000 in a metric ton.	**Kilometer** 1. 1,000 meters 2. It is 10 times larger than a hectometer. 3. It is equal to about .6 of a mile.
Line Graph 1. This type of graph compares two variables, with one variable plotted along a vertical axis and the other along a horizontal axis. 2. Shows a change over time with points that are connected by line segments. 3. In this type of graph each variable is plotted along an axis.	**Line Plot** 1. To make one, you start by drawing a horizontal number line. 2. Each value in the set is recorded by placing an "x" over the corresponding value. 3. By looking at the number of x/s over the number line on this type of chart, you can see how often the value occurred.
Linear Measurement 1. The measure of the length of an object. 2. Inches, feet, yards and miles are U.S. customary units of ___. 3. Meters, centimeters and kilometers are metric units of ___.	**Liter** 1. 1,000 milliliters. 2, 4 metric cups. 3. There are 1,000 of them in one kiloliter.
Mean 1. The average of all the numbers in a set of data. 2. In the following set of numbers, the ___ is 28: 20, 25, 25, 30, 40. 3. In the following set of numbers, the ___ is 41: 30, 30, 40, 44, 50, 52.	**Median** 1. The middle value in a set of numbers. 2. In the following set of numbers, the ___ is 28: 24, 25, 26, 28, 40, 41, 45. 2. In the following set of numbers, the ___ is 35: 22, 28, 32, 35, 38, 40, 42.
Meter 1. 1,000 millimeters 2. 100 centimeters 3. 10 decimeters	**Metric System** 1. This system of measurement is based on the decimal system. 2. This system of measurement is commonly used in the scientific community. 3. This system is based on the meter as a unit length, the kilogram as a unit of mass, and the liter as a unit of volume or capacity.

Mile 1. 5,200 feet. 2. 1,720 yards. 3. About 1.6 kilometers.	**Mode** 1. The value that occurs most frequently in a set of numbers. 2. In the following set of numbers, the ___ is 2: 1, 2, 2, 2, 2, 3, 4, 5, 5, 6, 7, 7, 8. 3. In the following set of numbers, the ___ is 8: 4, 5, 5, 6, 7, 7, 8, 8, 8, 8, 9.
Parallel 1. Lines that are in the same plane but never meet are called ___ lines. 2. ___ lines are at the same distance apart at all points. 3. Lines that are ___ to one another will never intersect.	**Perpendicular** 1. Lines that intersect at right angles are called ___ lines. 2. When ___ lines intersect, they form square corners at the point at which they meet. 3. Four right angles are formed by the intersection of lines of this type.
Plane 1. A 2-dimensional surface. 2. The set of all the points on an endless flat surface. 3. A flat surface that extends into infinity in all directions.	**Pound(s)** 1. 16 ounces. 2. There are 2,000 in a ton. 3. To convert them to kilograms, multiply by .45.
Range 1. The difference between the largest number and the smallest in a set of data. 2. In the following set of numbers, the ___ is 34: 32, 42, 44, 53, 54, 55, 66,. 3. In the following set of numbers, the ___ is 66: 51, 62, 74, 74, 85, 95, 97, 117.	**Scale** 1. The ___ of a drawing is the ratio of the size of the drawing to the actual size of the object. 2. A system of marks used for measuring. 3. If a map has a ___ of 1 inch per 2 miles, two points that are 3 inches apart would be 6 miles apart.
Square Foot (Feet) 1. There are 144 square inches in one. 2. There is 1 square yard in 9 of them. 3. There are 43,560 of them in an acre.	**Stem-and-Leaf Plot** 1. A method of organizing data in a way that shows its shape and distribution. 2. In a ___ the ones digits and the tens and greater digits are separated by a line. 3. The ones digits are the stems and the tens and greater digits are the leaves.

Data and Measurement Bingo

Tablespoon	Table
1. 3 teaspoons. 2. There are 16 of them in one cup. 3. There are 15 milliliters in one metric one.	1. An orderly arrangement of data, especially one in which the data are arranged in columns and rows. 2. A multiplication ___ is often used to help students learn multiplication facts. 3. Another word for this is "matrix."

Tally	Temperature
1. A recorded reckoning or account. 2. A mark used in recording a number of acts or objects, usually in series of five. 3. These marks are often grouped in fives, with sets of four vertical lines marked diagonally or horizontally by a fifth line.	1. The degree of hotness or coldness. 2. The U.S. Customary System uses the Fahrenheit scale to measure this. 3. The Metric System uses the Celsius scale to measure this.

Time	Variable(s)
1. We measure it in seconds, minutes, hours, days, weeks, months, and years. 2. Decades, centuries and millennia are also units of ___. 3. Chronometry is the science of the measurement of ___	1. It is something that is subject to change. 2. A quantity capable of assuming any set of values. 3. In the formula $a^2 + b^2 = c^2$, a, b, and c are ___.

Volume	Weight
1. The measure of the amount of space in a container. 2. To find the ___ of a cube, use the formula $V = s^3$, or side cubed. 3. Multiply length x width x height to find the ___ of a rectangular 3-dimensional figure.	1. Ounces, pounds and tons are units of ___ measurement in the U.S. Customary System. 2. The gram is the basic unit of mass in the Metric System. Mass is often referred to as ___ although they are not exactly the same. 3. Mass measures the *amount* of matter; ___ measures the force of attraction of earth acting on the object and is not constant.

X-Axis	Y-Axis
1. The horizontal axis of a two-dimensional rectangular coordinate system. 2. It is perpendicular to the y-axis. 3. The ___ plane goes from left to right and intersects with the y-axis	1. The vertical axis of a two-dimensional rectangular coordinate system. 2. It is perpendicular to the x-axis. 3. The ___ plane goes up and down and intersects with the x-axis

Data and Measurement Bingo

Data and Measurement Bingo

Square Foot (Feet)	28	Capacity	Bar Graph	Coordinate(s)
Metric System	Graph	Weight	Table	Scale
Area	Stem-and-Leaf Plot		Tally	Pound(s)
Celsius	Hour	Y-Axis	Line Plot	Parallel
Volume	Perpendicular	Mile	Frequency	Variable(s)

Data and Measurement Bingo

Celsius	Area	Line Graph	Range	Hectare(s)
Parallel	Table	Cubic Foot (Feet)	Hour	Mode
Gallon	Perpendicular		Kilogram	Y-Axis
Tablespoon	Volume	Stem-and-Leaf Plot	Time	Weight
Scale	Variable(s)	Mile	Metric System	Frequency

© Barbara M. Peller

Data and Measurement Bingo

Celsius	Y-Axis	Table	Line Plot	Area
Perpendicular	32 Fluid Ounces	Range	28	Data
Hour	Weight		Mode	55
Stem-and-Leaf Plot	Gallon	Volume	Tablespoon	Line Graph
Frequency	Metric System	Mile	Time	Kilogram

Data and Measurement Bingo: Card No. 3

Data and Measurement Bingo

Stem-and-Leaf Plot	Mode	Capacity	Metric System	Kilogram
Histogram	Century	28	Range	Area
Tally	Tablespoon		Coordinate(s)	Bar Graph
Y-Axis	Kilometer	Weight	Mile	Cubic Foot (Feel)
Data	Scale	Meter	Frequency	Pound(s)

Data and Measurement Bingo

Scale	Coordinate(s)	Hour	Cubic Foot (Feet)	Metric System
Histogram	Y-Axis	Fahrenheit	Hectare(s)	32 Fluid Ounces
Capacity	Pound(s)		Mean	Liter
Weight	Kilogram	Square Foot (Feet)	Time	Graph
Table	Mile	Area	Stem-and-Leaf Plot	Tally

Data and Measurement Bingo

55	Mode	Line Graph	Kilogram	Pound(s)
Line Plot	Hour	Graph	28	Area
Range	Circle Graph		Century	Hectare(s)
Mile	Volume	Time	Meter	Capacity
Parallel	Y-Axis	Square Foot (Feet)	Tally	Kilometer

Data and Measurement Bingo

Square Foot (Feet)	Mode	Mile	Mean	Table
Parallel	Hectare(s)	Perpendicular	Customary System	Histogram
Line Graph	Bar Graph		Kilogram	Century
Stem-and-Leaf Plot	Tablespoon	Fahrenheit	Celsius	Gallon
Mile	Metric System	Time	Meter	55

Data and Measurement Bingo

Tally	Mode	Customary System	Line Plot	Century
Histogram	Capacity	Range	Pound(s)	Cubic Foot (Feet)
Kilometer	Temperature		Kilogram	Coordinate(s)
Frequency	Stem-and-Leaf Plot	Celsius	Circle Graph	Tablespoon
Weight	Mile	Meter	Hour	Parallel

Data and Measurement Bingo

Kilogram	Table	Perpendicular	Kilometer	Metric System
Circle Graph	Hectare(s)	Tally	Hour	Mode
Data	Square Foot (Feet)		Graph	Customary System
32 Fluid Ounces	Weight	Volume	Mean	Liter
Tablespoon	Time	Fahrenheit	Celsius	Coordinate(s)

Data and Measurement Bingo

Celsius	Line Plot	Century	Range	Kilometer
Pound(s)	Cubic Foot (Feet)	28	Graph	Kilogram
Temperature	Mode		Bar Graph	Gallon
Volume	Weight	32 Fluid Ounces	Time	Data
Fahrenheit	Parallel	Line Graph	Scale	Tally

Data and Measurement Bingo

55	Mode	Hour	Graph	Parallel
Customary System	Data	Mean	Hectare(s)	28
Histogram	Kilogram		Line Graph	Perpendicular
Fahrenheit	Area	Time	Metric System	Celsius
Circle Graph	Mile	Square Foot (Feet)	Meter	Table

Data and Measurement Bingo

Table	Coordinate(s)	Area	Line Plot	Kilogram
Perpendicular	Parallel	Capacity	Meter	Graph
Square Foot (Feet)	Liter		Pound(s)	Range
Mile	Tablespoon	Hectare(s)	Celsius	Histogram
Mode	Customary System	Temperature	Circle Graph	Cubic Foot (Feet)

Data and Measurement Bingo

Graph	Coordinate(s)	55	Time	Pound(s)
Capacity	Customary System	Hectare(s)	Kilogram	Gallon
Line Plot	Cubic Foot (Feet)		Perpendicular	Liter
Tally	Time	Century	Temperature	Celsius
Mile	Weight	Meter	Square Foot (Feet)	Mean

Data and Measurement Bingo

Metric System	Hectare(s)	Hour	Kilogram	Circle Graph
Cubic Foot (Feet)	Square Foot (Feet)	Data	Graph	Mode
32 Fluid Ounces	Bar Graph		Line Graph	Fahrenheit
Weight	Time	Temperature	Century	55
Mile	Range	Gallon	Parallel	Tally

Data and Measurement Bingo

Mean	Hectare(s)	Hour	Table	Line Plot
55	Line Graph	28	Capacity	Circle Graph
Pound(s)	Square Foot (Feet)		Area	Mode
Mile	Data	Customary System	Time	Graph
Parallel	Tablespoon	Meter	Kilometer	Perpendicular

Data and Measurement Bingo

Century	Data	Customary System	Kilometer	Volume
Range	Gallon	Liter	Histogram	Bar Graph
32 Fluid Ounces	Coordinate(s)		Pound(s)	Perpendicular
Stem-and-Leaf Plot	Cubic Foot (Feet)	Mile	Mean	Celsius
Circle Graph	X-Axis	Meter	Tablespoon	Mode

Data and Measurement Bingo: Card No. 16

Data and Measurement Bingo

Fahrenheit	Median	Linear Measurement	Data	Metric System
Mean	Circle Graph	Time	Bar Graph	Liter
Hectare(s)	Tally		X-Axis	Customary System
Weight	Parallel	Celsius	Hour	Gallon
Volume	32 Fluid Ounces	Table	Line Plot	Coordinate(s)

Data and Measurement Bingo

Kilometer	Temperature	Cubic Foot (Feet)	Graph	Range
Mode	Fahrenheit	Volume	Pound(s)	Circle Graph
Hectare(s)	Gallon		Linear Measurement	Capacity
Weight	28	Time	Celsius	Line Graph
X-Axis	Data	Hour	Median	55

Data and Measurement Bingo

Pound(s)	55	Data	Customary System	Temperature
Mean	Line Plot	Mode	Table	Bar Graph
Median	Metric System		Graph	Area
Line Graph	X-Axis	Volume	Tablespoon	Linear Measurement
Capacity	Plane	Parallel	Tally	Meter

Data and Measurement Bingo

Temperature	Median	Line Plot	Data	Meter
Cubic Foot (Feet)	Perpendicular	Histogram	Plane	Range
Coordinate(s)	Liter		Stem-and-Leaf Plot	28
Scale	Weight	Frequency	Tablespoon	X-Axis
Y-Axis	Tally	Volume	Celsius	Linear Measurement

Data and Measurement Bingo

Mean	55	Histogram	Data	Scale
Coordinate(s)	Linear Measurement	Century	Customary System	Square Foot (Feet)
Gallon	Parallel		Median	Hour
Plane	Table	X-Axis	Weight	Tally
Stem-and-Leaf Plot	Volume	Meter	Fahrenheit	Tablespoon

Data and Measurement Bingo

Kilometer	Line Graph	Linear Measurement	Capacity	Graph
Range	Line Plot	Area	Customary System	32 Fluid Ounces
Cubic Foot (Feet)	Bar Graph		Square Foot (Feet)	Liter
X-Axis	Weight	Tablespoon	28	Metric System
Volume	Fahrenheit	Median	Gallon	Histogram

Data and Measurement Bingo

Century	Median	Table	Capacity	Meter
55	Temperature	Parallel	Mean	28
Line Graph	Graph		Frequency	Square Foot (Feet)
Gallon	Volume	X-Axis	Fahrenheit	Tablespoon
Scale	Weight	Tally	Plane	Linear Measurement

Data and Measurement Bingo

Century	Temperature	Metric System	Median	Customary System
Linear Measurement	Meter	Histogram	Range	Square Foot (Feet)
Liter	Kilometer		Graph	Gallon
Scale	Frequency	X-Axis	Fahrenheit	Coordinate(s)
Y-Axis	Stem-and-Leaf Plot	Volume	Line Plot	Weight

Data and Measurement Bingo

Stem-and-Leaf Plot	Histogram	Median	Hour	Linear Measurement
28	Weight	Mean	Century	32 Fluid Ounces
Coordinate(s)	Customary System		Frequency	X-Axis
Area	Scale	Variable(s)	Volume	Bar Graph
Meter	Metric System	Cubic Foot (Feet)	Circle Graph	Y-Axis

Data and Measurement Bingo

Linear Measurement	Median	Line Graph	Range	Kilometer
Plane	Line Plot	Customary System	Temperature	Century
Weight	Frequency		Bar Graph	Stem-and-Leaf Plot
Fahrenheit	Capacity	Scale	Volume	X-Axis
Liter	Circle Graph	Hour	Variable(s)	Y-Axis

Data and Measurement Bingo: Card No. 26

© Barbara M. Peller

Data and Measurement Bingo

Line Graph	Cubic Foot (Feet)	Median	Temperature	Perpendicular
Scale	Frequency	Mean	X-Axis	32 Fluid Ounces
Time	Weight		Volume	Stem-and-Leaf Plot
Kilometer	55	Histogram	Y-Axis	28
Circle Graph	Bar Graph	Linear Measurement	Area	Liter

Data and Measurement Bingo

Pound(s)	Temperature	Area	Median	Century
Perpendicular	Linear Measurement	Frequency	Range	Bar Graph
Weight	Gallon		Liter	Plane
Celsius	Kilometer	Parallel	Volume	X-Axis
Capacity	Hectare(s)	Circle Graph	Y-Axis	Scale

Data and Measurement Bingo

Linear Measurement	Temperature	Kilometer	Mean	Hectare(s)
Weight	Plane	Histogram	Liter	Area
Coordinate(s)	Frequency		32 Fluid Ounces	Median
Porpondicular	Scalo	Kilogram	Volume	X-Axis
Century	32 Fluid Ounces	Y-Axis	55	Variable(s)

Data and Measurement Bingo

Metric System	Median	Range	Kilogram	X-Axis
28	Temperature	Line Graph	Bar Graph	Graph
Variable(s)	32 Fluid Ounces		Liter	Histogram
Y-Axis	55	Capacity	Volume	Frequency
Scale	Table	Weight	Linear Measurement	Area

© Barbara M. Peller

www.ingramcontent.com/pod-product-compliance
Lightning Source LLC
Chambersburg PA
CBHW080720220326

41520CB00056B/7282